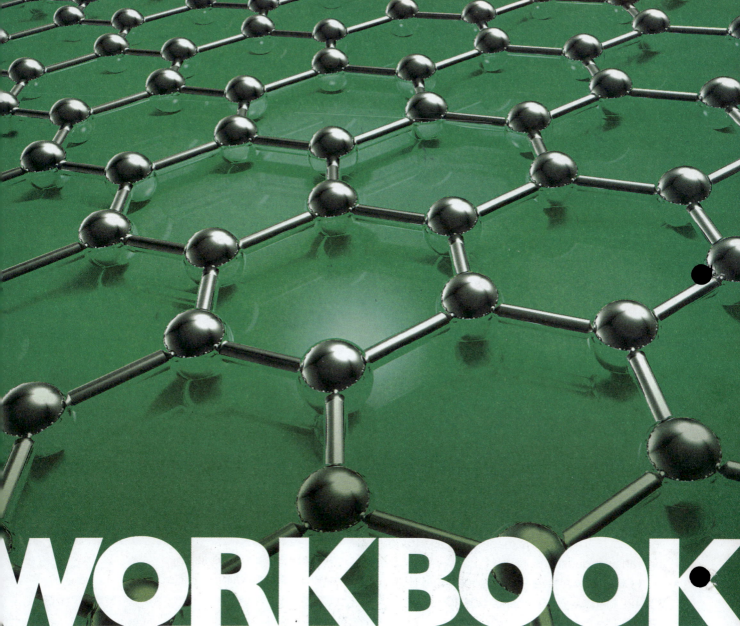

WORKBOOK

Chemistry

Physical chemistry 1

Alyn G. McFarland and Nora Henry

FOR THE NEW
2015
SPECIFICATIONS

PHILIP ALLAN FOR
HODDER
EDUCATION

LEARN MORE

Contents

WORKBOOK

①

This workbook will help you to prepare for the following exams:

- AQA Chemistry AS Paper 1: the exam is 1 hour 30 minutes long, worth 80 marks and 50% of your AS. The exam is made up of multiple choice, short, and long answer questions.
- AQA Chemistry AS Paper 2: the exam is 1 hour 30 minutes long, worth 80 marks and 50% of your AS. The exam is made up of multiple choice, short, and long answer questions.
- AQA Chemistry A-level Paper 1: the exam is 2 hours long, worth 105 marks and 35% of your A-level. The exam is made up of short and long answer questions.
- AQA Chemistry A-level Paper 2: the exam is 2 hours long, worth 105 marks and 35% of your A-level. The exam is made up of short and long answer questions.
- AQA Chemistry A-level Paper 3: the exam is 2 hours long, worth 90 marks and 30% of your A-level. The exam includes questions testing across the whole specification.

②

For each topic of each section there are:
- stimulus materials including key terms and concepts
- short-answer questions
- long-answer questions
- multiple-choice questions
- questions which test your mathematical skills
- space for you to write

③

Answering the questions will help you to build your skills and meet the assessment objectives (AO1) (knowledge and understanding), AO2 (application) and AO3 (analysis, interpretation and evaluation).

④

You still need to read your textbook and refer to your revision guides and lesson notes.

⑤

Marks available are indicated for all questions so that you can gauge the level of detail required in your answers.

⑥

Timings are given for the exam-style questions to make your practice as realistic as possible.

⑦

Answers are available at: www.hoddereducation.co.uk/workbookanswers

Physical chemistry

Topic 1 Atomic structure

Fundamental particles, mass number and isotopes

An atom consists of a nucleus containing protons of relative mass 1 and relative charge +1, and neutrons of relative mass 1 and relative charge 0 surrounded by electrons of relative mass 1/1840 and charge −1.

The **atomic (proton) number** (Z) is equal to the number of protons in the nucleus of an atom.

The **mass number** (A) is the total number of protons and neutrons in the nucleus of an atom.

Atoms are electrically neutral as they have equal numbers of protons and electrons. To calculate the numbers of each particle present in an atom use the equations:

number of protons = atomic number
number of neutrons = mass number − atomic number

When an atom loses electrons, it becomes a positive ion. When an atom gains electrons, it becomes a negative ion.

charge on an ion = number of protons − number of electrons

Isotopes of an element are atoms with the same number of protons but a different number of neutrons.

Hence isotopes of an element have the same atomic number but a different mass number.

A time-of-flight (TOF) mass spectrometer can be used to identify elements and to determine relative molecular mass. The five processes that occur in a TOF mass spectrometer are:

- electrospray ionisation to form gaseous positive ions
- acceleration to constant kinetic energy
- ion drift
- ion detection
- data analysis

The mass spectrum is a plot of relative abundance against mass to charge ratio (m/z). The abundance of each isotope is given by the height of the peak.

The relative atomic mass of an element can be calculated using

$$A_r = \frac{\Sigma(\text{mass of isotopes} \times \text{relative abundance})}{\Sigma(\text{relative abundance})}$$

where Σ represents the 'sum of' for all isotopes.

The relative molecular mass of the compound can be determined by looking for the peak with the largest m/z value — the molecular ion peak.

---

1 **How many electrons are there in an aluminium ion Al^{3+}? (AO2)** `1 mark`

   A 10

   B 13

   C 16

   D 27

2 Part of the mass spectrum for paracetamol is shown below. Which one of the following numbers is the molecular ion peak? (AO1) **1 mark**

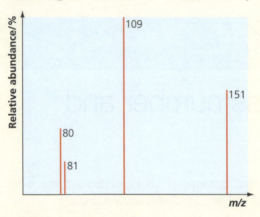

A 80

B 81

C 109

D 151

---

3 Complete the table to give information about some atoms and ions. (AO2) **5 marks**

Species	Number of protons	Number of neutrons	Number of electrons
K			
O^{2-}			
Fe^{2+}			
Ag^+			
Ba			

4 A sample of an element A was analysed and found to contain three isotopes with mass numbers and percentage abundances as shown in the table below.

Mass number	Percentage abundance
127	95.91
129	2.49
131	1.60

a i Define the term *isotope*. (AO1) **1 mark**

....

....

ii Use data from the table to calculate the relative atomic mass of A.
   Give your answer to 1 decimal place. (AO2) **3 marks**

....

....

....

....

iii The data from the table were obtained from a mass spectrum. Identify the ion responsible for the peak at 127. (AO2)

....

....

iv Explain how ions are accelerated, separated and detected in a time-of-flight (TOF) mass spectrometer. (AO1) **4 marks**

....

....

....

....

....

....

....

....

....

b Some data about a different element, B, were determined and recorded in the table below.

i Complete the table. (AO2) **1 mark**

Atomic number	83
Mass number	
Number of protons	
Number of neutrons	126
Number of electrons	

ii Identify element B. (AO2) **1 mark**

....

iii State the meaning of the term *mass number*. (AO1) **1 mark**

....

....

iv State the meaning of the term *atomic number*. (AO1) **1 mark**

....

....

Electron configuration

Electrons are arranged in energy levels that are subdivided into subshells, designated:

- s — holds up to 2 electrons
- p — holds up to 6 electrons
- d — holds up to 10 electrons

Subshells contain orbitals; two electrons can occupy each orbital. The subshells fill in the order:

$1s\ 2s\ 2p\ 3s\ 3p\ 4s\ 3d\ 4p$

When first series transition metal atoms form ions they lose their $4s$ electrons first.

The electron configurations of chromium and copper atoms are unusual:

Cr: $1s^2\ 2s^2\ 2p^6\ 3s^2\ 3p^6\ 3d^5\ 4s^1$ (not $3d^4\ 4s^2$)
Cu: $1s^2\ 2s^2\ 2p^6\ 3s^2\ 3p^6\ 3d^{10}\ 4s^1$ (not $3d^9\ 4s^2$)

The **first ionisation energy** is the energy required to remove 1 mole of electrons from 1 mole of gaseous atoms to form 1 mole of gaseous 1+ ions. The equation to represent this for an element X is:

$$X(g) \rightarrow X^+(g) + e^-$$

The equation to represent the second ionisation energy is:

$$X^+(g) \rightarrow X^{2+}(g) + e^-$$

Down a group the first ionisation energy *decreases* because there is an increase in atomic radius and increased shielding and hence less nuclear attraction of the outer electron.

Across a period first ionisation *increases* because there is an increase in nuclear charge, but shielding is similar as the electron is being removed from the same shell. As a result, there is a smaller atomic radius, as the outermost electron is held closer to the nucleus by the greater nuclear charge.

---

1 Write the electronic configuration of the following atoms and ions. (AO2)   **12 marks**

a H ....

b F ....

c Mg ....

d Al ....

e Fe ....

f Ni ....

g Cl^- ....

h N^{3-} ....

i V^{3+} ....

j Cu^{2+} ....

k Fe^{3+} ....

l S^{2-} ....

2 The graph shows the variation in first ionisation energy with atomic number.

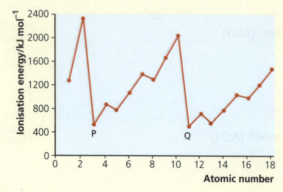

To which one of the following do the elements P and Q belong? (AO2)   **1 mark**

A alkali metals

B halogens

C noble gases

D transition metals

3 The first five successive ionisation energies, in kJ mol⁻¹, of an element Z are 799, 2420, 4000, 25 000 and 32 800.

 1 mark

Which group of the periodic table does element Z belong to? (AO2)

A group 1

B group 2

C group 3

D group 4

4 Write equations, including state symbols, for the reaction that occurs when the following ionisation energies are measured: (AO1)

4 marks

a first ionisation energy of calcium

b first ionisation energy of aluminium

c second ionisation energy of iron

d third ionisation energy of silicon

5 The bar chart below shows the first ionisation energies of the group 1 metals.

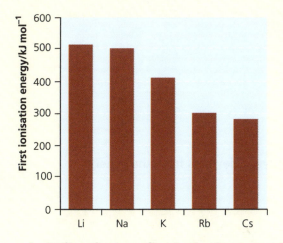

a i Define the term first ionisation energy. (AO1)

2 marks

ii State and explain the trend in first ionisation energy down group 1. (AO1)   `3 marks`

....

....

....

....

....

iii Write the full electron configuration of the Na^{2+} ion. (AO1)   `1 mark`

....

The bar chart below shows the first ionisation energies of the elements in period 2.

b i State and explain the general trend in ionisation energy across the period. (AO1)   `4 marks`

....

....

....

....

....

ii Explain why the first ionisation energy of oxygen deviates from the general trend. (AO2)   `2 marks`

....

....

....

iii Write an equation, including state symbols, for the reaction that occurs when the second ionisation energy of carbon is measured. (AO1)   `1 mark`

....

iv Give one reason why the second ionisation energy of carbon is lower than the second ionisation energy of boron. (AO2)   `1 mark`

....

....

Exam-style questions

1 Magnesium has three stable isotopes. Two of these are ^{25}Mg and ^{26}Mg.

a i Complete the table below to show the number of each type of fundamental particle in one atom of ^{25}Mg and ^{26}Mg.

2 marks

	Protons	Neutrons	Electrons
^{25}Mg			
^{26}Mg			

ii Use the periodic table to deduce the full electron configuration of magnesium.

1 mark

iii A sample of magnesium consisting of three isotopes has a relative atomic mass of 24.33. The table below gives the relative abundance of two of these isotopes.

Mass number of isotope	25	26
Relative abundance/%	10.11	11.29

Use this information to determine the relative abundance and hence the mass number of the third isotope.

Give your answer to the appropriate number of significant figures.

4 marks

iv State, in terms of fundamental particles, why the isotopes ^{25}Mg and ^{26}Mg have similar chemical reactions.

1 mark

v A time-of-flight (TOF) mass spectrometer can be used to determine the relative atomic mass of an element. The magnesium is ionised and the ions then accelerated.

Explain how and why the ions are accelerated.

2 marks

The first ionisation energy of magnesium is 738 kJ mol^{-1}.

b i Write an equation to show the process that occurs when the first ionisation energy of magnesium is measured. Include state symbols in your equation.

1 mark

ii Write an equation to show the process that occurs when the second ionisation energy of magnesium is measured. Include state symbols in your equation. `1 mark`

....

....

iii Explain why the second ionisation energy of magnesium is higher than the first ionisation energy of magnesium. `2 marks`

....

....

....

....

iv State and explain the trend in the first ionisation energies of the elements in group 2 from magnesium to barium. `3 marks`

....

....

....

....

....

Topic 2 Amount of substance

A balanced symbol equation gives the ratio of the number of moles of each reactant and product.

One mole of a substance is the mass in grams that contains the Avogadro constant of particles (6.02×10^{23}).

For an ideal gas:

$$pV = nRT$$

where p = pressure of a gas (Pa), V = volume (m^3), n = amount (in moles), R = gas constant ($JK^{-1}mol^{-1}$) and T = temperature (K).

An **empirical formula** is the simplest ratio of the atoms in a compound. Empirical formulae are used for giant covalent compounds and ionic compounds.

Heating a hydrated compound to constant mass produces an anhydrous compound. The simplest ratio of the number of moles of the anhydrous compound to the number of moles of water determines the degree of hydration.

$$\text{pertcentage yield} = \frac{\text{actual yield}}{\text{theoretical yield}} \times 100$$

$$\text{atom economy} = \frac{M_r \text{ of useful atoms in the products}}{\text{total } M_r \text{ of all reactant atoms}} \times 100$$

A titration is carried out using a pipette and burette and two solutions are mixed to determine the exact volume of one solution required to react with an exact volume of the other solution. Acid–base titrations are common where one solution is a base (an alkali) and the other is an acid.

An indicator is used to determine the exact point when the acid has neutralised the base or vice versa. Two common indicators are phenolphthalein and methyl orange.

The preparation (including rinsings) and accurate use of the apparatus are important in obtaining reliable results.

The volume of solution added from the burette is called the **titre**. One rough and two accurate titrations are carried out and the average titre is the average of the two *accurate* titres.

Calculations are carried out based on the average titre to determine:
- concentration
- number of moles
- mass
- M_r
- percentage purity
- degree of hydration
- identity of unknown elements

---

1 **What is the colour change observed when hydrochloric acid is added to sodium hydroxide solution containing phenolphthalein indicator? (AO1)**   `1 mark`

A red to yellow

B yellow to red

C pink to colourless

D colourless to pink

2 The results of a titration are shown below.

	Titration 1	Titration 2	Titration 3	Titration 4
Final burette reading/cm³	25.40	23.10	24.00	22.70
Initial burette reading/cm³	3.10	0.30	1.00	0.00
Titre/cm³	22.30	22.80	23.00	22.70

Which results should be used to calculate the average titre? (AO3)   1 mark

A all of them

B 1 and 2

C 2 and 3

D 2 and 4

3 Calcium phosphate can be represented by the formula $Ca_3(PO_4)_x$. 6.21 g of calcium phosphate contain 1.24 g of phosphorus and 2.56 g of oxygen. Determine the empirical formula of calcium phosphate and hence the value of x. (AO2)   4 marks

4 24.70 cm³ of a solution of potassium hydroxide of concentration 0.422 mol dm⁻³ react completely with 25.00 cm³ of a solution of sulfuric acid. What is the concentration of the sulfuric acid in mol dm⁻³? Give your answer to 3 significant figures. (AO2)   3 marks

5 Five indigestion tablets containing calcium carbonate with a total mass of 2.50 g were added to 50.0 cm³ of 1.00 mol dm⁻³ hydrochloric acid.

$$CaCO_3 + 2HCl \rightarrow CaCl_2 + CO_2 + H_2O$$

After reaction the solution was transferred to a 250 cm³ volumetric flask and the volume made up to 250 cm³ using deionised water. 25.0 cm³ samples of the solution were titrated against 0.126 mol dm⁻³ sodium hydroxide solution.

$$NaOH + HCl \rightarrow NaCl + H_2O$$

The average titre was 11.30 cm³.

a Calculate the mass of calcium carbonate (in grams) in the five tablets. Give your answer to 3 significant figures. (AO2)   5 marks

....
....
....
....
....
....
....
....
....
....

b Calculate the percentage of calcium carbonate in the tablets. (AO2)   1 mark

....
....
....
....
....

6 A sample of hydrated sodium carbonate has the formula $Na_2CO_3.xH_2O$. 3.48 g of this sample were dissolved in deionised water and the solution transferred to a volumetric flask. The volume was made up to 250 cm³ using deionised water. A 25.0 cm³ sample of this solution was titrated against 0.105 mol dm⁻³ hydrochloric acid and the average titre was found to be 26.50 cm³.

$$Na_2CO_3 + 2HCl \rightarrow 2NaCl + CO_2 + H_2O$$

Determine the value of x. (AO2)   5 marks

....

....

....

....

....

....

....

....

....

7 Sodium azide (Na_3N) decomposes on heating to form sodium and nitrogen gas.

$$2Na_3N \rightarrow 6Na + N_2$$

1.35 g of sodium nitride were heated.

a Calculate the mass of sodium formed on complete decomposition of the sodium nitride. Give your answer to 3 significant figures. (AO2)   3 marks

....

....

....

....

....

b Calculate the volume of nitrogen gas (in cm³) produced on complete decomposition of 1.35 g of sodium nitride at 320 K and 110 kPa. The gas constant $R = 8.31\,J\,K^{-1}\,mol^{-1}$. Give your answer to 3 significant figures. (AO2)   3 marks

....

....

....

....

....

....

....

8 1,3,5-tribromobenzene ($C_6H_3Br_3$) may be synthesised from benzene according to the equation:

$C_6H_6 + 3Br_2 \rightarrow C_6H_3Br_3 + 3HBr$

a Calculate the percentage atom economy of this reaction if 1,3,5-tribromobenzene is the desirable product. Give your answer to 3 significant figures. (AO2)   **2 marks**

b 5.50 cm³ of benzene (density = 0.877 g cm⁻³) were used with an excess of bromine and 11.58 g of 1,3,5-tribromobenzene were obtained. Calculate the percentage yield. Give your answer to 3 significant figures. (AO2)   **4 marks**

9 2.48 g of hydrated copper(II) chloride ($CuCl_2.xH_2O$) were heated to constant mass. 1.96 g of anhydrous copper(II) chloride were obtained.

Determine the value of x in $CuCl_2.xH_2O$. (AO2)   **4 marks**

Eye test

1 In what directions did Joe have to look?

_____

_____

2 What did Joe have to notice about the circles?

_____

_____

3 Which letters could Joe read?

_____

_____

4 How did the optician use the strange glasses that Joe wore?

_____

_____

5 What did the optician decide about Joe's eyes?

_____

_____

6 On a separate piece of paper, write about a time when you visited the optician or the dentist. Why did you have to go? What happened?

Notes for teachers
Help the children to read the passage slowly and carefully, ensuring that they understand the story. Discuss the questions with them and encourage them to work out their answers orally before putting anything down on paper. Do they remember to write in complete sentences, using appropriate punctuation? Note that to answer question 3 the pupils should describe the positions of the letters rather than attempting to list the actual letters that Joe would have seen.

Eye test

1   What was the optician's room like?

_____

_____

2   Why did the optician look so carefully into Joe's eyes?

_____

_____

3   Why did Joe have to read the letters?

_____

_____

4   Why did he have to wear strange glasses?

_____

_____

5   How did the optician know that Joe needed some new glasses?

_____

_____

6   On a separate piece of paper, write about a time when you visited an optician or a dentist. Why did you have to go? Where did you have to go? What was special about the room you visited? What happened?

Notes for teachers
Help the children to read the passage slowly and carefully, ensuring that they understand the story. Discuss the questions with them and encourage them to write their answers in complete sentences, using appropriate punctuation. Can the pupils produce a detailed answer to question 6?

My teacher's car

My teacher's car
is a brand new car
with clean and shining wheels.
"I am so proud,"
she says out loud.
I don't know how she feels.

My teacher's car
won't get too far.
It's only got three wheels.
It did have four,
Not any more!
I wonder how she feels.

My teacher's car
won't get too far.
It's only got two wheels.
It did have three,
then one broke free.
I wonder how she feels.

My teacher's car
won't get too far.
It's only got one wheel.
It did have two,
but now not new.
I wonder how she feels.

My teacher's car's
a useless car
'cos it has got no wheels.
She's got a bike,
or she could hike.
I wonder how she feels.

My teacher's car

1 What is the teacher's car like to start with?

_____

_____

2 How does the car change in verse two of the poem?

_____

_____

3 What happened to a wheel in verse three?

_____

_____

4 How many wheels has the car got in verse four?

_____

_____

5 What is the car like at the end of the poem?

_____

_____

6 How do you think the teacher feels about the car at the end?

_____

_____

Notes for teachers
Help the children to read the poem, ensuring that they understand the simple sequence of events. Discuss the questions with them and encourage them to work out their answers orally before putting anything down on paper. Do they remember to write in complete sentences, using appropriate punctuation?

My teacher's car

1 How does the teacher feel in verse one of the poem?

_____

_____

2 Why does she feel that way?

_____

_____

3 What happens to the car?

_____

_____

4 What could be causing the problems with the wheels?

_____

_____

5 How do you think the teacher feels by the end of the poem?

_____

_____

6 How can the teacher manage without a car?

_____

_____

Notes for teachers
Help the children to read the poem, ensuring that they understand the simple sequence of events. Discuss the questions with them and encourage them to work out their answers orally before putting anything down on paper. Do they remember to write in complete sentences, using appropriate punctuation?

My teacher's car

Read the poem 'My teacher's car'. The poem tells a story. Write the story in your own words, describing what happens. Try to explain how the teacher feels and how she or he is going to manage for transport in future.

_____

_____

_____

_____

_____

_____

_____

_____

_____

_____

_____

_____

_____

_____

Notes for teachers
Give the children the poem to read by themselves. Are they able to write out the simple story, perhaps adding their own details to expand upon the sequence of events? Do they remember to write in complete sentences, using appropriate punctuation?

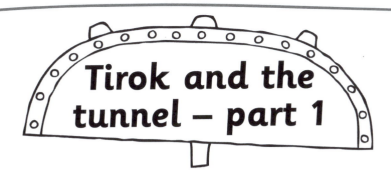

Tirok and the tunnel – part 1

The text below is from 'Tirok', a story set in the future, about a boy and his alien friends on a journey to Earth. At this point in the story they are beginning to uncover a mystery. On their spacecraft there is a tunnel not shown on any plan of the ship.

The tunnel shaped walkway, down which they quietly crept, appeared to be much like all the others on the Endeavour. Yet somehow it felt quite different. In other parts of the vessel there was the hum of quietly running engines, the chatter of voices in many different languages and the intermittent messages heard over the ship's communication system. Here however, on this newly found route, there was just an eerie silence.

As they followed the curved way forward, the sense of courage they had started out with gradually ebbed away.

Jen was the first to voice what they were all beginning to think when she whispered, "Maybe this wasn't such a smart idea, it's rather eerie along here."

"I think you could be right," agreed Sote, also in a whisper. "Let's just go on for a few more minutes, and if we don't find anything interesting we'll turn back."

They continued to follow the winding path. It took them past several unusual gleaming silvery doors. At each, they paused hoping there would be a way of seeing inside. Outside the third one, Tirok noticed that Jen had the sort of look on her face that you might expect a person to have if they were concentrating very hard on a particularly tricky piece of school work. The three friends were a little disappointed that all the doors were very solid and very closed.

Shortly after passing the fourth door, they arrived at a very strange junction. To be more exact they nearly fell down it. The path quite suddenly dropped about a metre, before splitting in three different directions. Luckily Tirok, always observant, had spotted it just in time. His arm suddenly shot out to bar the way to the others, preventing them from taking a nasty tumble.

"That was close!" said Jen.

By Judy Richardson

Tirok and the tunnel – part 1

Ring the correct answers to the questions.

1 How many children are in the story?

> One   two   three   four

2 How did the children speak?

> They sang.   They whispered.   They shouted.   They laughed.

3 Where were the children walking?

> On a hill.   Along a street.   By a river.   In a tunnel.

Answer each of the next questions with a full sentence.

4 What were the doors like?

_____

_____

5 How many doors did they pass?

_____

_____

6 What did Tirok spot just in time?

_____

_____

Notes for teachers

Read the passage through with the children – do they understand what is happening in the story? When tackling the first three questions, encourage pupils to look at all the possible answers before deciding which one is correct.

 Andrew Brodie: More Improving Comprehension for Ages 8–9 © A&C Black, Bloomsbury Publishing 2012

Tirok and the tunnel – part 1

Ring the correct answers to the questions.

1 Which word is nearest in meaning to 'winding'?

curved   straight   vertical   horizontal

2 Which word is nearest in meaning to 'followed'?

disappointed   chased   after   pursued

3 Who was the first person to speak?

Tirok   Jen   Sote   the captain

Answer the questions below using full sentences.

4 Who suggested turning back?

_____

_____

5 Why were the children disappointed?

_____

_____

6 What was strange about the junction?

_____

_____

Notes for teachers
Ask the children to read the story out loud, taking it in turns to read – do they understand what is happening in the story? When tackling the first three questions, encourage pupils to look at all the possible answers before deciding which one is correct. They may find a dictionary or thesaurus helpful.

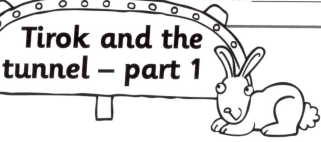

Tirok and the tunnel – part 1

Ring the correct answers to the questions.

1   Which word is nearest in meaning to 'intermittent'?

occasional   continuous   constant   regular

2   Which word is nearest in meaning to 'gleaming'?

shiny   locked   double   glowing

3   Why did Tirok's arm shoot out?

To hit an   To block   To catch   To hit the
animal.   the way.   Jen.   light switch.

Answer the questions below using full sentences.

4   Why was Tirok good at spotting any danger?

_____

_____

5   Describe the walk that the children have taken.

_____

_____

_____

_____

_____

_____

Notes for teachers
Ask the children to read the passage to themselves before reading it out loud to each other in pairs. Encourage them to consider the words in questions 1 and 2 in the context of the passage – which of the options would work best in the context?

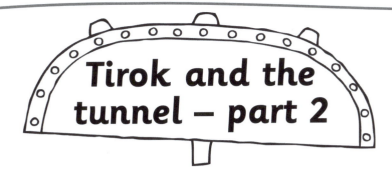

Tirok and the tunnel – part 2

"What a strange arrangement," commented Sote. "There is a drop then we would have a choice of three ways. I really don't think it would be wise to go any further."

"Well not today anyway," agreed Jen.

For the first time since they had started, Tirok spoke. His voice was very quiet and tinged with anxiety. "I have a nasty feeling that we would be in real trouble if we were found here." (By this he didn't mean the sort of trouble you might get into if you were caught in the wrong classroom during the lunch break, he meant the sort of trouble spelt D – A – N – G – E – R!)

"We should go back quickly and quietly and talk about it properly when we know we are safe."

No one argued with such an obviously sensible idea, and no one spoke as they headed back the way they came. Again they were aware of the complete silence, broken only by the hushed sound of their soft-soled shoes on the glistening floor beneath their feet. It was not a moment too soon for any of them when they emerged onto Floating Walkway 17 on deck 42.

Feeling very relieved, the three adventurers quickly found the café, ordered drinks and chocolate muffins and began to talk about the tunnel. Back in the usual busy atmosphere of this part of the ship the children began to feel that perhaps they had imagined the danger. Perhaps they had misread the map and had been down a corridor that was, after all, quite ordinary. Perhaps the silver doors were just entrances to work rooms. Or perhaps they had imagined the whole thing.

Never, however, do three people all imagine the same thing at the same time. Little could they possibly have imagined the extent of the danger they had come close to in that walkway and just how perilous it would be to go past the three way junction they had reached that day.

Feeling quite cheered, Tirok, Jen and Sote arranged to meet up again the next day to do some more exploring. That was when the trouble would really begin.

By Judy Richardson

Tirok and the tunnel – part 2

Ring the correct answers to the questions.

1   Which word is nearest in meaning to 'nasty'?

   mean   horrible   smelly   ugly

2   Which word is nearest in meaning to 'trouble'?

   strife   prison   crime   danger

3   What did the children eat?

   muffins   nothing   chocolate   biscuits

Answer each of the next questions with a full sentence.

4   What were the children's shoes like?

   _____

   _____

5   What deck did they arrive at?

   _____

   _____

6   What did the children decide to do the next day?

   _____

   _____

Notes for teachers
This passage follows on from part 1 on page 13. Read the passage through with the children – do they understand what is happening in the story? Encourage them to consider the words in questions 1 and 2 in the context of the passage – which of the options would work best in the context?

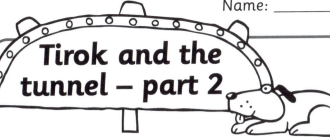

Tirok and the tunnel – part 2

Ring the correct answers to the questions.

1   Which word is nearest in meaning to 'drop'?

     fall   break   clumsy   release

2   Which word is opposite in meaning to 'sensible'?

     clever   stupid   sensitive   sensation

Answer the questions below using full sentences.

3   Where did the three children first feel safe again?

   _____

   _____

4   What did they buy at the café?

   _____

   _____

5   How did their feelings change when they were at the café?

   _____

   _____

6   What might happen the next day?

   _____

   _____

   _____

Notes for teachers

This passage follows on from part 1 on page 13. Ask the children to read the story out loud, taking it in turns to read – do they understand what is happening in the story? Encourage them to consider the words in questions 1 and 2 in the context of the passage – which of the options would work best in the context?

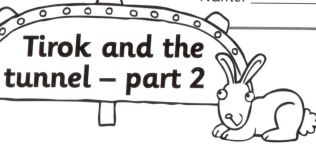

Tirok and the tunnel – part 2

Ring the correct answers to the questions.

1 Which word is nearest in meaning to 'caught'?

      kidnapped   snatched   found   arrested

2 Which word is nearest in meaning to 'perilous'?

      dangerous   tortuous   scandalous   mysterious

Answer the questions below using full sentences.

3 How can we tell that the space-ship is very large?

_____

_____

4 What was the area near the café like?

_____

_____

5 Why did their feelings change when they were at the café?

_____

_____

6 What could happen next? On a separate sheet of paper write a continuation of the story.

Notes for teachers

This passage follows on from part 1 on page 13. Ask the children to work in pairs, reading the story out loud to each other. Encourage them to consider the words in questions 1 and 2 in the context of the passage – which of the options would work best in the context?

Sydney Opera House

The Sydney Opera House is one of the most famous buildings in the world. It is situated alongside the harbour in the city of Sydney in Australia.

Although it is in Australia, the Opera House was designed by an architect from Denmark. Jom Utzon produced his design in 1957 but the building wasn't finished until sixteen years later.

Many performing arts events take place in the Sydney Opera House, not just opera! People can visit the Opera House to watch plays, ballet or other performances. They can also listen to music performed by orchestras or pop singers.

Over seven million people come to the Sydney Opera House each year just to see the amazing building. It has several different sections including a concert hall, three theatres and two other large halls. All of these are contained in the giant shapes that look like shells or sails.

Another famous landmark nearby is the Sydney Harbour Bridge, which was completed in 1932.

Sydney Opera House

1   In which country is the Sydney Opera House?

_____

_____

2   What does an architect do?

_____

_____

3   What country was the architect of the Sydney Opera House from?

_____

_____

4   How many years did it take to build the Sydney Opera House?

_____

_____

5   How many people visit the Sydney Opera House each year?

_____

_____

6   What other famous building is near to the Sydney Opera House?

_____

_____

Notes for teachers
Help the children to read the passage, ensuring that they understand that it is a piece of non-fiction writing. They may find it helpful to see other photographs of the building, perhaps by using computer software to show satellite images. Discuss the questions with them and encourage them to write their answers in full sentences.

Sydney Opera House

1   Where is the Sydney Opera House?

_____

_____

2   When was the Sydney Opera House first designed?

_____

_____

3   In what year was the Sydney Opera House completed?

_____

_____

4   What performing arts events could you see at the Sydney Opera House?

_____

_____

5   How many theatres are contained within the Sydney Opera House?

_____

_____

6   How much older than the Sydney Opera House is the Sydney Harbour Bridge?

_____

_____

Notes for teachers
Help the children to read the passage, ensuring that they understand that it is a piece of non-fiction writing. They may find it helpful to see other photographs of the building, perhaps by using computer software to show satellite images. Discuss the questions with them and encourage them to write their answers in full sentences.

Sydney Opera House

1 Why do you think the Sydney Opera House is so famous?

_____

_____

2 What was the nationality of the architect of the Sydney Opera House?

_____

_____

3 Why might people visit the Sydney Opera House?

_____

_____

4 Where might you go in this country to see performing arts events?

_____

_____

5 Do you know any famous buildings in this country? Find out about one of them and write some key points about it.

_____

_____

_____

_____

_____

_____

Notes for teachers
Ensure that the children understand that the passage is a piece of non-fiction writing. They may find it helpful to see other photographs of the building, perhaps by using computer software to show satellite images. Discuss the questions with them and encourage them to write their answers in full sentences. Talk about ideas that they could use to answer question 6. They may need some help finding information about their chosen building.

Andrew Brodie: More Improving Comprehension for Ages 8–9 © A&C Black, Bloomsbury Publishing 2012

Golden Eagle

General information

The golden eagle is a large bird of prey. Golden eagles live in mountainous areas though some also live by the coast. Most of the golden eagles in the United Kingdom live in the Highlands of Scotland. They are not easy to see as they tend to keep away from humans.

Food

Golden eagles mainly eat rabbits, hares, grouse and other small birds. They will also take small lambs. They swoop down on their prey, travelling at a speed of up to ninety miles per hour.

Identification

The golden eagle is very large and has a wing-span of approximately two metres. The adult is dark brown with a golden tinge to the head. Eagles can be seen soaring in the sky but they sometimes fly with slow wing beats.

Nesting

Golden eagles' nests are known as eyries. The eyrie is made from sticks and is usually located on the edge of a cliff. The eagle usually lays two eggs, which are dull white in colour. The eggs take about six weeks to hatch. The eaglets begin to learn to fly about eleven or twelve weeks after hatching.

Habitat

Mountains, high forests, sea coasts.

New neighbour (2)

Use full sentences to answer the questions below.

1   Why did Jason place the Morse Code print out on his window-sill?

_____

_____

2   Why did Jason get so tired?

_____

_____

3   What woke Jason?

_____

_____

4   What was the first message that Jason received?

_____

_____

5   How would Jason flash out his message in Morse Code? Show the message in dots and dashes.

_____

_____

_____

_____

_____

Notes for teachers
Ask the children to take turns to read the paragraphs of the text. Do they understand the sequence of events? Talk about the questions and encourage the children to compose sentences orally before they write them down. Ensure that the children can see clearly the print out of Morse Code.

New neighbour (2)

Name: _____   Date: _____

Use full sentences to answer the questions below.

1 How did Jason nearly miss the message from across the road?

_____

_____

2 How would Sam flash out the final message in Morse Code? Show the message in dots and dashes.

_____

_____

_____

_____

_____

3 What message would you send in Morse Code and who would you send it to? On the back of the sheet, write the message in Morse Code dots and dashes.

_____

_____

4 What do you think could happen next in the story?

_____

_____

_____

_____

Notes for teachers

Ask the children to take turns to read the paragraphs of the text. Do they understand the sequence of events? Talk about the questions and encourage the children to compose sentences orally before they write them down. Ensure that the children can see clearly the print out of Morse Code.